THE BUSINESS SUCCESS GUIDE TO BEEKEEPING

Essential Techniques, Strategies For Profitable Hive Management, Sustainable Practices, Marketing Your Honey Products, For Beginners And Experts

RICHMOND HAMILL

© 2024 [RICHMOND HAMILL]. All rights reserved.

Except for brief quotations included in critical reviews and certain other noncommercial uses allowed by copyright law, no part of this book may be reproduced, distributed, or transmitted in any form or by any means, including photocopying, recording, or other electronic or mechanical methods, without the publisher's prior written permission.

Disclaimer

The information presented in this book is based on the author's personal knowledge and understanding of livestock management. The author is not affiliated with any association, company, business, or individual in the livestock industry. All content is provided for informational purposes only and should not be considered as professional advice. Readers are encouraged to seek professional guidance and conduct their own research before making any decisions based on the information contained in this book. The author and publisher disclaim any liability for any adverse effects or consequences resulting from the use of the information contained herein.

TABLE OF CONTENTS

CHAPTER ONE 21

Introduction To Beekeeping 21

Understanding The Importance Of Bees 21

History Of Beekeeping 22

Benefits Of Beekeeping 23

Basic Beekeeping Equipment 25

Overview Of Beekeeping Responsibilities 26

CHAPTER TWO 29

Getting Started With Beekeeping 29

Choosing The Right Location For Your Hives 29

    Assessing Environmental Factors 29

    Considering Safety and Accessibility 30

    Evaluating Forage Availability 30

Selecting The Right Beekeeping Equipment ... 31

    Essential Hive Components 31

- Protective Gear and Tools 32
- Additional Equipment for Hive Management 32
- Understanding Local Beekeeping Regulations 33
  - Researching Legal Requirements 33
  - Complying with Health and Safety Standards 34
  - Joining Local Beekeeping Associations 34
- How To Source Your First Bees 35
  - Choosing Between Package Bees and Nucleus Colonies 35
  - Selecting a Reputable Supplier 36
  - Transporting and Introducing Bees to the Hive 37
- Setting Up Your First Hive 38
  - Assembling and Positioning the Hive 38

Installing the Bees 38

Monitoring and Maintenance 39

CHAPTER THREE 41

Understanding Bee Biology 41

Anatomy Of A Honeybee 41

    Head: 41

    Thorax: 42

    Abdomen: 42

The Life Cycle Of A Bee 43

    Egg: 43

    Larva: 43

    Pupa: 44

    Adult: 44

Roles Within The Hive: Queen, Worker, And Drone 45

    Queen: 45

    Worker: 45

Drone: 46

How Bees Make Honey 46

Nectar Collection: 46

Processing: 47

Storage: 47

Bee Communication And Behavior 48

Communication: 48

Behavior: 48

CHAPTER FOUR 51

Managing Your Hive 51

Inspecting Your Hive: What To Look For 51

Identifying And Addressing Common Hive Issues 52

Feeding Your Bees: When And What To Feed . 54

Swarm Prevention And Management 55

Seasonal Hive Management Tips 56

CHAPTER FIVE 59

Bee Health And Disease Management 59

Common Bee Diseases And Pests 59

Identifying Signs Of Disease 60

Preventative Health Measures 62

Treatment Options For Common Issues 63

Maintaining A Healthy Hive 64

CHAPTER SIX 67

Harvesting Honey 67

When And How To Harvest Honey 67

   Timing is Everything 67

   Checking the Hive 67

   Handling Frames 68

Tools And Equipment For Harvesting 69

   Essential Tools 69

     1.   Hive Tool: 69

     2.   Bee Brush: 69

3. Extractor: 69

4. Uncapping Knife: 70

5. Strainers and Filters: 70

Setting Up 70

Extracting Honey: Methods And Techniques .. 71

Manual Extraction 71

1. Uncapping: 71

2. Spinning: 71

3. Draining: 71

Hot Knife Method 72

Storing Your Harvested Honey 72

Proper Storage Conditions 72

Choosing Containers 73

Labeling and Dating 73

Legal Considerations For Selling Honey 74

Understanding Regulations 74

Labeling Requirements...................................74

Business Licensing 75

Insurance and Liability 75

CHAPTER SEVEN..77

Expanding Your Apiary 77

When To Add More Hives 77

Techniques For Splitting Hives 79

Managing Multiple Hives...............................81

Scaling Up: From Hobby To Small Business ...82

Tips For Successful Apiary Expansion............84

CHAPTER EIGHT 87

Pollination And Bees 87

The Role Of Bees In Pollination......................87

Benefits Of Pollination For Gardens And Crops
....89

Attracting Bees To Your Garden 90

The Impact Of Beekeeping On Local Ecosystems 92

Supporting Wild Bee Populations 94

CHAPTER NINE 97

Advanced Beekeeping Techniques 97

Queen Rearing And Replacement 97

    Introduction to Queen Rearing 97

    Methods of Queen Rearing 97

    Benefits of Queen Rearing 98

    Practical Steps in Queen Rearing 98

        1. Preparing Queen Cups: 98

        2. Grafting Technique: 99

        3. Nurturing Queens: 99

        4. Introducing New Queens: 99

Advanced Hive Designs And Configurations 100

    Introduction to Advanced Hive Designs 100

    Types of Advanced Hive Designs 100

Benefits of Advanced Hive Designs.............101

Practical Implementation of Advanced Hive Designs 101

1. Choosing the Right Design:...................101

2. Constructing the Hive:..........................101

3. Managing the Hive:...............................102

Organic Beekeeping Practices 102

Introduction to Organic Beekeeping...........102

Principles of Organic Beekeeping...............103

Benefits of Organic Beekeeping..................103

Implementing Organic Beekeeping Practices 104

1. Hive Management:................................104

2. Pest and Disease Control:.....................104

3. Nutrition and Habitat:.........................104

Overwintering Strategies For Bees 104

Introduction to Overwintering....................104

Methods of Overwintering 105

Benefits of Effective Overwintering 105

Practical Steps in Overwintering 106

    1.   Assessing Hive Strength: 106

    2.   Insulating Hives: 106

    3.   Monitoring and Maintenance: 106

Utilizing Bee Products Beyond Honey 107

  Introduction to Bee Products 107

  Types of Bee Products 107

  Benefits of Diversifying Bee Product Use ... 107

  Practical Applications of Bee Products 108

    1.   Harvesting and Processing: 108

    2.   Marketing and Sales: 108

    3.   Regulatory Considerations: 108

CHAPTER TEN 109

  Troubleshooting And Support 109

Common Problems And Their Solutions 109

When To Seek Professional Help....................111

Resources For Continued Learning................112

Building A Beekeeping Community...............114

Preparing For Long-Term Beekeeping Success
....115

Frequently Asked A Question And Answers ..119

CONCLUSION 127

THE END...131

ABOUT THIS BOOK

"Beekeeping" is an essential guide for anyone interested in the fascinating world of apiculture. It begins with an introduction that highlights the crucial role of bees in our ecosystem, their historical significance, and the numerous benefits of beekeeping. The introductory section also familiarizes readers with the fundamental equipment needed and provides an overview of the responsibilities that come with managing hives.

Starting a beekeeping venture requires careful planning and knowledge, which is thoroughly covered in the next part of the book. Readers will find detailed advice on selecting the perfect location for their hives, choosing the right equipment, and understanding local regulations. This section also guides novices on sourcing their first bees and setting up their initial hive, ensuring a solid foundation for a successful beekeeping experience.

An in-depth understanding of bee biology is crucial for any beekeeper. This book delves into the anatomy of honeybees, their life cycle, and their distinct roles within the hive. It explains how bees produce honey and the intricate ways they communicate and behave, providing readers with a comprehensive insight into the inner workings of a bee colony.

Managing a hive effectively is key to maintaining a healthy and productive bee population. This book offers practical tips on hive inspections, identifying and addressing common issues, and feeding bees appropriately. It also covers swarm prevention and seasonal management practices, ensuring that beekeepers can keep their hives thriving throughout the year.

Bee health and disease management are critical components of beekeeping. This book provides valuable information on common bee diseases and pests, signs of illness, and preventative health measures. It also explores treatment options and strategies for maintaining a healthy hive, equipping beekeepers with the knowledge to protect their colonies from harm.

Harvesting honey is one of the most rewarding aspects of beekeeping. This guide details the best times and methods for harvesting honey, the necessary tools and equipment, and various extraction techniques. It also covers storage practices and legal considerations for those interested in selling their honey, providing a complete roadmap from hive to market.

For those looking to expand their beekeeping operations, this book offers expert advice on adding more hives, splitting existing hives, and managing multiple colonies. It includes tips on scaling up from a hobby to a small business, ensuring that beekeepers can grow their apiaries successfully and sustainably.

The role of bees in pollination and their benefits to gardens and crops are also discussed in detail. This book explains how to attract bees to your garden, the positive impact of beekeeping on local ecosystems, and ways to support wild bee populations, highlighting the broader environmental significance of beekeeping.

Advanced beekeeping techniques are covered for those looking to deepen their expertise. This section explores queen rearing and replacement, advanced hive designs, organic practices, and overwintering strategies. It also discusses the use of bee products beyond honey, offering a wealth of knowledge for experienced beekeepers.

Finally, This book addresses common problems and solutions, guides on when to seek professional help, and lists resources for continued learning. It emphasizes the importance of building a beekeeping community and preparing for long-term success, making it an indispensable resource for both novice and seasoned beekeepers.

CHAPTER ONE

Introduction To Beekeeping

Understanding The Importance Of Bees

Bees are crucial to our ecosystem due to their role in pollination. They are responsible for pollinating approximately one-third of the food crops we consume, including fruits, vegetables, and nuts. Without bees, many of these plants would struggle to reproduce, leading to reduced food supplies and biodiversity. Additionally, bees help maintain the health of ecosystems by supporting the growth of trees, flowers, and other plants, which serve as food and shelter for various creatures.

Pollination occurs when bees transfer pollen from the male part of a flower to the female part, facilitating fertilization. This process is not only vital for food production but also for the survival of wild plants. Bees, particularly honeybees, are efficient pollinators because they visit numerous flowers each day. By understanding and valuing the role of bees, beekeepers contribute to the preservation of these essential pollinators and the stability of our environment.

History Of Beekeeping

Beekeeping, or apiculture, has a rich history dating back thousands of years. The earliest evidence of beekeeping is found in ancient Egyptian art and records, where beekeepers are depicted using simple hives and collecting honey. The practice spread to ancient Greece and Rome, where it was refined and documented. Greek philosophers like Aristotle wrote about the behavior and biology of bees, while Roman poets praised the sweetness of honey.

In medieval Europe, beekeeping became an integral part of monastic life, with monks maintaining apiaries for honey and beeswax, which were used for food, medicine, and candles. The development of movable-frame hives in the 19th century revolutionized beekeeping by allowing beekeepers to inspect hives without destroying them, improving honey production and bee health.

Modern beekeeping has evolved with advancements in technology and science. Today, beekeepers use sophisticated tools and techniques to manage colonies, monitor bee health, and maximize honey yields. Despite these changes, the fundamental principles of beekeeping remain rooted in ancient practices.

Benefits Of Beekeeping

Beekeeping offers numerous benefits beyond honey production. One of the most significant advantages is the contribution to pollination and the health of the environment. By maintaining bee colonies, beekeepers support the pollination of crops and wild plants, ensuring a diverse and stable ecosystem.

Beekeeping also provides economic benefits. Honey, beeswax, propolis, royal jelly, and other bee products are valuable commodities. Beekeepers can sell these products locally or online, creating additional income streams. Moreover, beekeeping can promote local agriculture by enhancing crop yields through effective pollination.

For many, beekeeping is a rewarding hobby that offers personal satisfaction and a sense of connection to nature. It provides an opportunity to learn about bee behavior, biology, and ecology. The process of tending to hives and observing bees can be therapeutic and educational, fostering a deeper appreciation for these incredible insects and their role in our world.

Basic Beekeeping Equipment

To start beekeeping, you'll need some essential equipment. The most important piece is the hive, which consists of several components. A typical hive includes a bottom board, brood boxes, honey supers, frames, a queen excluder, an inner cover, and a top cover. The bottom board serves as the hive's foundation, while the brood boxes house the queen and brood (eggs, larvae, and pupae). Honey supers are placed above the brood boxes for honey storage.

Frames fit inside the boxes and provide a structure for bees to build their comb. A queen excluder is a mesh panel placed between the brood boxes and honey supers to keep the queen out of the honey area. The inner cover and top cover protect the hive from weather and pests.

Protective gear is essential for beekeepers. A bee suit or jacket, gloves, and a veil will protect you from stings. A smoker is used to calm bees during hive inspections, making them easier to manage. A hive tool is indispensable for prying apart hive components and scraping off excess wax and propolis.

Overview Of Beekeeping Responsibilities

Beekeeping involves various responsibilities to ensure the health and productivity of your bee colonies. Regular hive inspections are crucial to monitor the condition of the bees and the hive. During inspections, check for signs of disease, pests, and the queen's health. Look for eggs, larvae, and capped brood, as well as honey and pollen stores.

Feeding bees may be necessary during times of scarcity, such as early spring or late fall. Sugar syrup, pollen patties, and fondant are common supplements.

Beekeepers must also manage hive space by adding or removing frames and supers as needed to prevent overcrowding or swarming.

Pest and disease management is a critical aspect of beekeeping. Varroa mites, small hive beetles, and wax moths are common pests that can harm bee colonies. Regular monitoring and treatment, using chemical or natural methods, help control these threats.

Harvesting honey is one of the most rewarding aspects of beekeeping. Once the honey supers are full and the honey is capped, use an extractor to remove honey from the frames. Strain and bottle the honey for storage or sale.

By understanding these responsibilities and using the proper equipment, you can maintain healthy and productive bee colonies, contributing to the environment and enjoying the many benefits of beekeeping.

CHAPTER TWO

Getting Started With Beekeeping

Choosing The Right Location For Your Hives

Assessing Environmental Factors

When starting beekeeping, choosing the right location for your hives is crucial for the health and productivity of your bees. The ideal location should provide ample sunlight, as bees thrive in warm environments. Select a spot that receives morning sunlight to help the bees start their day early, but avoid areas with extreme afternoon heat. Proximity to water is also essential, as bees need water for cooling their hive and diluting honey. Ensure there is a nearby source of clean water, or place a shallow container with pebbles near the hive for the bees to access.

Considering Safety and Accessibility

Safety is another important consideration. Place your hives in a location where they won't be disturbed by pets, children, or livestock. Bees can become defensive if their hive is threatened, so it's best to keep them in a secluded area away from high-traffic zones. Additionally, ensure the site is easily accessible for you to perform regular hive inspections and maintenance. Clear, level ground will make it easier to maneuver equipment and work with the hives.

Evaluating Forage Availability

The availability of forage is critical for the bees' food supply. Choose a location with diverse flowering plants that provide nectar and pollen throughout the seasons. Bees require a steady food source, so having a variety of flowers, trees, and crops nearby will support their foraging needs. Avoid areas heavily treated with pesticides, as these chemicals can harm the bees. If necessary, consider planting a bee-friendly garden with a range of nectar-rich plants to enhance forage availability.

Selecting The Right Beekeeping Equipment

Essential Hive Components

Starting with the right beekeeping equipment is essential for successful hive management. The basic hive components include the hive stand, bottom board, brood boxes, honey supers, inner cover, and outer cover. The hive stand keeps the hive off the ground, protecting it from moisture and pests. The bottom board serves as the hive's floor, while the brood boxes are where the queen lays her eggs and the colony raises its young. Honey supers are placed above the brood boxes to store honey, and the inner and outer covers protect the hive from weather conditions.

Protective Gear and Tools

Protective gear is crucial to ensure your safety while working with bees. A beekeeping suit, gloves, and a veil will protect you from stings. Choose light-colored, smooth fabrics to avoid attracting bees. Essential tools include a hive tool for prying apart hive components, a smoker to calm the bees during inspections, and a bee brush to gently move bees off frames. Having the right equipment will make hive management easier and safer.

Additional Equipment for Hive Management
In addition to the basic components and protective gear, several other pieces of equipment can aid in hive management.

A queen excluder, placed between the brood boxes and honey supers, prevents the queen from laying eggs in the honey storage area. Frame grips help lift heavy frames without damaging them. An uncapping knife or fork is used to remove wax cappings from the honeycomb during extraction. Investing in a good quality bee feeder can help supplement the bees' diet during times of scarcity.

Understanding Local Beekeeping Regulations

Researching Legal Requirements

Before starting beekeeping, it's essential to understand local regulations and legal requirements. Regulations vary by location, so research the specific rules in your area. Contact your local agriculture department or beekeeping association to get information on permits, hive registration, and zoning laws.

Some regions may have restrictions on the number of hives you can keep, the distance from property lines, or the types of hives allowed.

Complying with Health and Safety Standards

Health and safety standards are also important to ensure the well-being of your bees and the community. Regulations may require regular hive inspections to monitor for diseases and pests. Some areas mandate beekeepers to maintain a certain distance between hives and public areas, such as schools or parks. Understanding and complying with these standards will help prevent issues with neighbors and local authorities, and promote responsible beekeeping practices.

Joining Local Beekeeping Associations

Joining a local beekeeping association can provide valuable support and resources. Associations often offer educational programs, mentorship opportunities, and access to bulk purchasing of equipment and bees. They can also keep you informed about changes in regulations and best practices. Networking with other beekeepers allows you to share experiences, gain insights, and stay updated on local beekeeping conditions.

How To Source Your First Bees

Choosing Between Package Bees and Nucleus Colonies

Sourcing your first bees can be done in a few ways, the most common being by purchasing package bees or nucleus colonies (NUCs). Package bees consist of a queen and several thousand worker bees shipped in a screened box. This option is often more affordable and available from many suppliers. However, it may take some time for the bees to build a comb and establish their hive. Nucs, on the other hand, are small, established colonies with a laying queen, brood, and honey stores. They are typically more expensive but provide a quicker start to beekeeping as the colony is already functioning.

Selecting a Reputable Supplier

Choosing a reputable supplier is crucial for obtaining healthy bees. Look for suppliers with good reviews and recommendations from other beekeepers. Local suppliers are often preferable as their bees are acclimated to the regional climate. Ensure the supplier adheres to health standards and provides documentation on the bees' health status, including treatments for pests like Varroa mites. Visiting the supplier's apiary can give you a better idea of their practices and the quality of their bees.

Transporting and Introducing Bees to the Hive

Transporting bees requires careful handling to minimize stress and injury. If purchasing package bees, keep them cool and shaded during transport. Once at the hive location, mist the package lightly with water to keep the bees hydrated. To introduce them to the hive, gently shake the bees into the prepared hive, ensuring the queen is safely introduced using a queen cage. For nucs, transfer the frames from the nuc box into the hive, maintaining the same order to preserve the colony's organization. Allow the bees to settle and provide them with a sugar syrup feeder to support them as they adjust to their new home.

Setting Up Your First Hive

Assembling and Positioning the Hive

Setting up your first hive involves assembling and positioning it correctly. Start by placing the hive stand in the chosen location, ensuring it is level and stable. Assemble the hive components, starting with the bottom board, followed by the brood boxes, frames, and supers. Use a queen excluder if desired. Position the inner cover and outer cover on top to protect the hive from weather. Ensure the entrance reducer is in place to regulate the hive's airflow and protect against pests.

Installing the Bees

Once the hive is assembled, it's time to install the bees. If using package bees, remove the queen cage and place it between the frames in the brood box. Shake the worker bees into the hive, ensuring they have access to the queen. For nucleus colonies, transfer the frames from the nuc box into the brood box, maintaining the same order. Close the hive and allow the bees to settle. Provide a feeder with sugar syrup to support the bees as they establish their new home.

Monitoring and Maintenance

Regular monitoring and maintenance are essential for the health of your hive. Conduct inspections every 7-10 days to check for signs of disease, pests, and the queen's activity. Look for brood patterns, honey stores, and overall bee behavior. Address any issues promptly, such as treating Varroa mites or supplementing with pollen patties if forage is scarce. Keep the hive clean and free from debris, and replace any damaged components. Proper hive management will ensure a thriving and productive bee colony.

CHAPTER THREE

Understanding Bee Biology

Anatomy Of A Honeybee

The honeybee's anatomy is vital for its diverse roles and functions within the hive. Understanding this anatomy can simplify the beekeeping process. A honeybee's body is divided into three primary sections: the head, thorax, and abdomen.

Head: The head houses crucial sensory organs and tools. The compound eyes on either side of the head allow bees to detect movement and color, essential for foraging. The antennae are multi-functional, serving as touch sensors and chemical detectors to help the bee navigate and communicate. Additionally, the mandibles are powerful jaws used for grooming, hive construction, and processing food.

Thorax: The thorax is the middle segment and is crucial for mobility. It contains three pairs of legs and two pairs of wings. The legs are adapted for different functions such as pollen collection, cleaning, and hive maintenance. The wings are vital for flight, enabling bees to forage for nectar and pollen, as well as to transfer between hives and flowers.

Abdomen: The abdomen houses the digestive and reproductive systems. It contains the honey stomach, where nectar is stored before being transferred to the hive. For female bees, it also houses the ovaries and the stinger. The stinger is a defense mechanism, primarily used by worker bees when the hive is threatened.

To observe these anatomical features in action, carefully inspect a bee under a magnifying glass. Notice the segmented body, the functional arrangement of legs and wings, and the differences between worker and queen bees.

The Life Cycle Of A Bee

The life cycle of a bee is a fascinating process with four distinct stages: egg, larva, pupa, and adult. Each stage plays a critical role in the development of a bee.

Egg: The life cycle begins when the queen bee lays eggs in the hive's cells. The eggs are tiny, oval-shaped, and appear white. Over three days, they develop into larvae. This stage is critical as the eggs need consistent warmth and care from the worker bees.

Larva: After hatching, the larvae are fed a diet of royal jelly, honey, and pollen. They grow rapidly and shed their skin multiple times. This stage is marked by a significant increase in size as the larvae prepare for the next stage.

Pupa: The larvae then spin themselves into a cocoon and enter the pupal stage. Inside the cocoon, the bee undergoes metamorphosis, where its body structures are reorganized to become an adult bee. This stage lasts about two weeks.

Adult: The mature bee emerges from the pupa with fully developed wings and body. Worker bees typically start foraging and hive maintenance, while drones' primary role is mating, and queens focus on egg-laying. Each bee's role is crucial to the hive's survival and efficiency.

Understanding the life cycle helps in managing hive health and anticipating bee needs. Regular hive inspections can help monitor these stages and ensure proper care.

Roles Within The Hive: Queen, Worker, And Drone

Each bee in the hive has a specific role, contributing to the hive's overall function and productivity.

Queen: The queen is the hive's reproductive member. She is the largest bee, with a longer abdomen that contains ovaries for egg-laying. Her primary function is to lay eggs, which she does at a rate of up to 2,000 per day during peak season. The queen also releases pheromones that regulate hive behavior and maintain social harmony.

Worker: Worker bees are female but are not reproductive. They are responsible for various tasks including foraging for nectar and pollen, tending to larvae, cleaning the hive, and defending it from intruders. Workers have specialized body parts such as pollen baskets on their hind legs and wax glands to build honeycombs.

Drone: Drones are male bees and their primary role is to mate with a queen. They have larger bodies and are equipped with large eyes for spotting queens during mating flights. Drones do not participate in nectar gathering or hive maintenance. They are generally expelled from the hive in the winter when resources are scarce.

By understanding these roles, beekeepers can manage their colonies more effectively, ensuring that each type of bee performs its function efficiently.

How Bees Make Honey

Honey production is a complex process involving nectar collection, processing, and storage.

Nectar Collection: Worker bees forage for nectar from flowers. They use their proboscis (a long, tube-like tongue) to suck nectar into their honey stomach, a special organ designed for this purpose. On their journey, they collect pollen, which is essential for the hive's diet.

Processing: Upon returning to the hive, the foraging bees regurgitate the nectar into the cells of the honeycomb. Other worker bees then help process the nectar by adding enzymes and reducing its water content by fanning their wings. This evaporation process thickens the nectar into honey.

Storage: Once the honey reaches the right consistency, bees cap the honeycomb cells with beeswax to preserve it. This stored honey serves as a food reserve for the hive, especially during winter when foraging is not possible.

For beekeepers, understanding honey production helps in managing the hive's nectar resources and knowing when to harvest honey. Ensure that you leave enough honey for the bees' needs before taking any for yourself.

Bee Communication And Behavior

Bees have complex communication methods and behaviors that ensure the hive's smooth operation.

Communication: Bees communicate primarily through pheromones, which are chemical signals used to relay information about hive conditions, queen status, and danger. For example, a queen's pheromones signal her presence and regulate worker bee behavior. Additionally, bees use the "waggle dance" to convey information about the location of food sources. This dance involves specific movements and directions that help other bees find the nectar.

Behavior: Bees exhibit various behaviors that contribute to hive efficiency. Foraging bees use scent trails and visual cues to find flowers. They also perform tasks in a division of labor based on their age and role within the hive.

For example, younger bees clean and care for the larvae, while older bees handle foraging and hive defense.

Observing bee behavior provides insight into hive health and efficiency. Regularly monitoring bee activities and communication patterns can help you address issues like disease or foraging difficulties, ensuring a thriving hive.

CHAPTER FOUR

Managing Your Hive

Inspecting Your Hive: What To Look For

Inspecting your hive is a crucial part of beekeeping, ensuring the health and productivity of your colony. Begin your inspection by checking the external condition of the hive. Look for signs of wear or damage on the hive body and cover, ensuring that the structure is intact and secure. Check that the entrance is clear and unobstructed to allow easy access for the bees.

When opening the hive, use a hive tool to gently pry apart the frames. Start by inspecting the top brood box. Look for the queen and assess her laying pattern.

A healthy queen should have a consistent pattern of eggs and larvae. Check for signs of brood diseases, such as deformed larvae or capped broods with unusual colors. Next, inspect the frames for honey and pollen stores. Adequate honey stores are crucial, especially before winter.

Ensure that there is enough space in the hive for the bees to expand. If the hive is overcrowded, consider adding a box. Finally, monitor the behavior of the bees. Aggressive behavior or excessive absconding may indicate problems. Regular inspections will help you catch issues early and keep your hives thriving.

Identifying And Addressing Common Hive Issues

Understanding and addressing common hive issues is essential for maintaining a healthy beekeeping environment. One common issue is hive beetles, which can damage the comb and brood.

Look for small, dark beetles in the hive and on the frames. To manage hive beetles, consider using beetle traps or a screened bottom board. Keeping the hive clean and well-ventilated also helps.

Another issue to watch for is Varroa mites, which can weaken and kill bees. Check for mites by using a sticky board or an alcohol wash. If you find a high infestation, treat it with appropriate miticides or natural treatments like essential oils. It's crucial to follow treatment guidelines to avoid harming the bees.

Foulbrood diseases, such as American Foulbrood (AFB) or European Foulbrood (EFB), can decimate a colony. Symptoms include dead larvae with a foul smell or a sunken appearance. If foulbrood is suspected, it is often best to consult a local beekeeping expert and consider removing and burning infected frames to prevent spreading.

Feeding Your Bees: When And What To Feed

Feeding your bees is important to ensure they have enough resources, especially during times when natural forage is scarce. In spring, feed a sugar syrup solution (1:1 ratio of sugar to water) to stimulate brood rearing and hive buildup. During fall, switch to a thicker syrup (2:1 ratio) to provide ample storage for winter.

When feeding, use a hive feeder to avoid introducing contaminants. The feeder should be placed between the hive body and the inner cover or as a top feeder. Ensure that the bees have access to clean water at all times, as it is crucial for their hydration and overall health.

If you live in an area with long winters, you may need to provide supplemental feeding to ensure the bees have enough food.

Consider using fondant or candy boards during this period. Monitor the hive's food stores regularly to adjust feeding as needed. Proper feeding helps maintain a strong, healthy colony.

Swarm Prevention And Management

Swarm prevention and management are vital to keep your hive productive and stable. Swarming occurs when a colony becomes overcrowded or the queen's pheromones are no longer effective. To prevent swarming, ensure that the hive has adequate space by adding supers as needed. Regular inspections and frame manipulation can help manage hive expansion and prevent crowding.

If you notice signs of swarm preparation, such as queen cells or increased bee activity, take action immediately. Perform a split by dividing the hive and relocating part of the colony to a new hive body.

This process helps reduce overcrowding and gives the bees space to thrive. Additionally, ensure that you regularly inspect for swarm cells and remove them if detected early.

In case your colony does swarm, have a plan in place for managing the swarm. If you find the swarm, you can capture it using a bee vac or by shaking the cluster into a new hive box. Be sure to place the new hive in a suitable location and provide adequate care to the captured bees to help them establish a new colony.

Seasonal Hive Management Tips

Managing your hive through the seasons is essential for maintaining colony health and productivity. In spring, focus on building up the hive by ensuring the bees have ample space and resources. Monitor the brood pattern and food stores, and be prepared to add additional boxes as needed.

During summer, regular inspections are critical to check for swarm preparations and to ensure that the hive is not overcrowded. Keep an eye on honey production and manage pests to prevent infestations. Provide extra space for honey storage and keep the hive cool and well-ventilated during hot weather.

In the fall, start preparing the hive for winter by reducing the entrance size to help keep the warmth in. Ensure the hive has sufficient honey stores for winter and consider adding a windbreak or insulation if needed. In winter, check occasionally for signs of problems like moisture or inadequate food stores, but minimize hive disturbances to avoid stressing the bees.

By following these seasonal management tips, you can help ensure that your hive remains healthy and productive throughout the year.

CHAPTER FIVE

Bee Health And Disease Management

Common Bee Diseases And Pests

Understanding common bee diseases and pests is crucial for maintaining a healthy hive. Some of the most prevalent diseases include American Foulbrood (AFB), European Foulbrood (EFB), and Nosema. American Foulbrood is a bacterial infection that affects the brood (larvae and pupae) of bees, leading to a foul smell and the death of infected larvae. European Foulbrood, also a bacterial disease, causes larvae to die before they mature, but it generally has a less pronounced odor compared to AFB. Nosema, a fungal infection, impacts the digestive system of adult bees, leading to decreased lifespan and productivity.

In addition to diseases, bees are also susceptible to pests such as Varroa mites and wax moths. Varroa mites are parasitic pests that attach to adult bees and their brood, feeding on their bodily fluids and weakening the colony. Wax moths are another common pest that can damage the hive by laying eggs in the comb, where larvae then feed on the wax and honey, leading to significant damage.

To manage these issues, it is important to regularly inspect your hive and be vigilant for any signs of disease or pests. This proactive approach will help you address problems before they escalate.

Identifying Signs Of Disease

Detecting signs of disease early can save your hive from severe damage. For American Foulbrood, watch for symptoms such as dead brood in various stages of development, a dark and sunken appearance of the brood cells, and a foul, sour odor emanating from the hive. The presence of AFB can be confirmed by examining the brood pattern and the characteristic sticky, ropy substance found in dead larvae.

European Foulbrood can be identified by observing larvae that are yellowish to brown, with a watery, slimy appearance. Unlike AFB, EFB-affected brood often has a less pungent odor. Infected larvae may be found in the brood cells in varying stages of decomposition.

Nosema infection in adult bees can manifest as dysentery, where bees are seen with feces on their bodies or outside the hive. Other signs include reduced hive activity and poor overall bee health.

Regular inspections and monitoring for these symptoms are key to early detection and effective management of bee diseases.

Preventative Health Measures

Preventing disease and pest problems start with maintaining a clean and well-managed hive. Ensure that the hive is in a location with good ventilation to prevent moisture buildup, which can contribute to the growth of pathogens. Regularly remove any old or damaged combs and replace them with new, clean ones to reduce the risk of disease and pests.

Feeding your bees with high-quality, pathogen-free food can also help boost their immune system and overall health. Providing supplemental feed during periods of scarcity or when natural food sources are insufficient can prevent malnutrition and associated health issues.

Implementing a regular hive inspection schedule is another crucial preventive measure. By routinely checking for signs of disease and pests, you can address issues promptly and reduce the likelihood of major outbreaks.

Treatment Options For Common Issues

When a disease or pest problem is detected, prompt treatment is essential. For American Foulbrood, the most effective treatment is to destroy the infected hive equipment and replace it with new materials. In some cases, antibiotics may be used, but they should be administered under the guidance of a bee health specialist.

European Foulbrood can be managed with a combination of better hive management practices and the use of antibiotics or other treatments as recommended by a professional. Increasing the strength of the colony through requeening or combining weaker colonies can also help the hive recover.

Varroa mites can be controlled using a variety of methods, including chemical treatments such as miticides, or organic options like powdered sugar dusting or essential oil treatments. Regular monitoring and treatment are necessary to keep mite populations under control.

For wax moth infestations, removing and destroying infested combs and ensuring proper hive hygiene can help manage and prevent further damage.

Maintaining A Healthy Hive

Maintaining a healthy hive involves consistent management practices and regular monitoring. Ensure that the hive environment is conducive to bee health by providing adequate ventilation, proper insulation, and protection from extreme weather conditions. Regularly inspect the hive for signs of disease, pests, and overall hive condition.

Providing your bees with a balanced diet, including access to clean water and high-quality feed, will support their immune system and overall vitality. Additionally, maintaining good hive hygiene by removing old or damaged equipment and ensuring that the hive is free from mold and debris will contribute to a healthier bee colony.

By implementing these practices, you can help ensure that your hive remains productive and resilient, reducing the likelihood of disease and pest issues and supporting the long-term success of your beekeeping efforts.

CHAPTER SIX

Harvesting Honey

When And How To Harvest Honey

Timing is Everything

The best time to harvest honey is when the honey is fully ripened. This usually happens in late summer or early fall, just before the bees enter their winter phase. You'll know it's time when the honeycomb cells are capped with beeswax. Capping indicates that the honey has been cured and is at the right moisture level for storage. Uncapped honey is too watery and can ferment, leading to spoilage.

Checking the Hive
To determine if honey is ready for harvest, inspect the hive. Look for signs of excess honey in the supers (the boxes placed above the brood boxes where bees

store honey). A good indication is when the bees are not adding more honey and are instead preparing for winter. Use a bee suit, gloves, and a smoker to calm the bees while you work. Gently remove the supers, and use a bee brush to sweep away any bees before bringing the frames inside for extraction.

Handling Frames

When harvesting, be careful not to damage the frames or disturb the bees more than necessary. It's advisable to work in a well-ventilated area to prevent overheating. If you're new to beekeeping, consider starting with a small batch of frames to get the hang of the process before moving on to larger quantities.

Tools And Equipment For Harvesting

Essential Tools

1.   Hive Tool: This is a multipurpose tool used for prying apart hive components, scraping off excess wax, and separating honey frames from the hive. It's essential for accessing the frames and removing them without damaging the hive.

2.   Bee Brush: A gentle brush is used to remove bees from the frames without harming them. It helps keep the honey clean and free from bee debris.

3.   Extractor: This is a device used to spin the honey out of the frames. Extractors come in manual and electric varieties. Manual extractors require hand cranking, while electric extractors are powered by a motor and are more efficient for larger operations.

4.   **Uncapping Knife:** This knife is used to remove the wax caps from the honeycomb cells. It can be heated to make the process easier, ensuring a clean extraction.

5.   **Strainers and Filters:** After extraction, honey needs to be filtered to remove any residual wax, bee parts, or debris. Strainers and filters come in various grades and sizes depending on the level of filtration required.

Setting Up

Before starting, make sure all your tools and equipment are clean and sanitized. This prevents contamination of the honey. Set up your extraction area in a clean, dry location where you can manage the mess and keep everything organized. Having a dedicated space for honey harvesting helps maintain hygiene and efficiency.

Extracting Honey: Methods And Techniques

Manual Extraction

1. Uncapping: Begin by using the uncapping knife to carefully remove the wax caps from the honey cells. This exposes the honey and allows it to flow out. Do this over a clean container to catch any drips and wax pieces.

2. Spinning: Place the uncapped frames in the extractor. If using a manual extractor, turn the handle slowly to spin the frames. The centrifugal force will pull the honey out of the cells and onto the sides of the extractor. In an electric extractor, adjust the speed settings as necessary to control the extraction process.

3. Draining: After spinning, allow the honey to drain from the extractor into a clean bucket or

container. Some extractors have a honey gate for this purpose, making it easier to direct the honey into jars or bottles.

Hot Knife Method

For small quantities, you can use a heated knife to cut off the cappings. The heat helps melt the wax, making it easier to remove. After uncapping, place the frames in an extractor or let the honey drain into a container. This method is effective but can be labor-intensive for larger harvests.

Storing Your Harvested Honey

Proper Storage Conditions

Honey should be stored in a cool, dry place away from direct sunlight. Ideal storage temperatures are between 50-70°F (10-21°C). Honey is hygroscopic, meaning it absorbs moisture from the air, which can lead to fermentation if not stored correctly.

Choosing Containers

Store honey in airtight containers to maintain its quality. Glass jars are a popular choice as they don't interact with the honey and allow for easy monitoring of the product. Plastic containers are also used but should be food-grade and free from chemical contaminants.

Labeling and Dating

Label each container with the date of harvest and any other relevant information, such as the type of honey or the hive's location. Proper labeling helps with tracking and ensuring freshness. Honey can remain good for years if stored properly, but it's helpful to have a record of when it was harvested.

Legal Considerations For Selling Honey

Understanding Regulations

Before selling honey, familiarize yourself with local regulations regarding honey production and sales. Regulations may vary by region but often include guidelines on labeling, food safety, and business permits. Some areas require honey to be inspected and approved by health authorities before it can be sold.

Labeling Requirements

Labels must include information such as the product name (e.g., "Raw Honey"), the net weight, the producer's contact details, and any pertinent allergen information. Some regions also require nutritional information and the country of origin. Ensure your labels are compliant with local standards to avoid legal issues.

Business Licensing

If you plan to sell honey commercially, you may need a business license or permit. Check with your local business regulatory agency to determine the necessary steps for legal operation. This might include registering your business, obtaining a sales tax permit, and adhering to health department guidelines.

Insurance and Liability

Consider obtaining insurance to protect against potential liabilities, such as customer complaints or product claims. Liability insurance can provide peace of mind and help manage risks associated with selling food products.

CHAPTER SEVEN

Expanding Your Apiary

When To Add More Hives

Adding more hives to your apiary is a crucial step in expanding your beekeeping operation. One of the primary indicators that it's time to increase the number of hives is the observation of hive congestion. If the current hive is overcrowded, with bees frequently clustering outside or within the hive, it's a signal that the hive is outgrowing its space. This typically happens in the spring and summer when the colony is in its peak activity and production period. Another key indicator is when the hive is producing excess honey that the bees can't store due to space constraints, suggesting they need more room to spread out.

A second critical moment for adding hives is when you notice that a colony has strong, healthy, and active characteristics, such as a high brood pattern and consistent honey production. Healthy, thriving colonies can quickly become too large for their current hive. Monitoring the population density and productivity will help you decide the right time to add more hives. Additionally, if you observe multiple frames of brood, it is a good practice to consider adding another hive to prevent swarming and to maintain hive health.

Finally, seasonal changes and environmental factors play a role in deciding when to add more hives. Spring and early summer are the best times to expand because the bees are entering their productive phase. Adding hives during these times allows the colonies to build up their numbers and resources before the winter months. Avoid expanding your apiary during late summer or fall when the bees are preparing for winter, as this can disrupt their preparation process.

Techniques For Splitting Hives

Splitting hives is a common method used to expand your apiary and prevent swarming. The process involves dividing a strong, healthy colony into two or more separate colonies, each with its queen. This not only helps manage hive overcrowding but also creates additional colonies that can grow and produce honey. The first technique is the classic split method, where you take one frame of brood, one frame of honey, and a few frames of empty comb from an existing hive and place them into a new hive body. Ensure the new hive has a queen cell or a mated queen.

Another technique is the nuc split, where you create a nucleus colony (nuc) from the parent hive. A nuc typically consists of a smaller hive box containing frames with brood, honey, and a mated queen. To perform this split, choose a strong hive and set aside a few frames of brood, honey, and pollen, transferring them into a nuc box. This nuc can then be placed in a new location, and the original hive will continue with the remaining frames and a new queen or queen cell.

The walk-away split is a less hands-on technique where you simply divide the colony into two parts, moving one part to a new location and leaving the other in place. After a few days, you inspect both hives to ensure each has a queen. This method relies on the bees' natural behavior to resolve the issue of overcrowding and can be less stressful for the bees than other methods.

Managing Multiple Hives

Managing multiple hives requires organization and careful monitoring to ensure the health and productivity of each colony. Start by keeping detailed records of each hive's status, including the date of inspections, hive conditions, and any treatments applied. This helps in tracking the progress and needs of each hive. You might use a simple logbook or digital beekeeping software for this purpose.

Regular inspections are crucial. Check each hive every 1-2 weeks during the active season to monitor the queen's health, brood patterns, and honey production. During these inspections, look for signs of diseases, pests, or signs that the hive might be preparing to swarm. Be prepared to take action, such as re-queening or treating pests, based on your observations.

Another important aspect is managing resources. Ensure that each hive has adequate space by adding supers or extra frames as needed. Be prepared for seasonal changes by providing necessary treatments and feeding as required. During winter, make sure the hives are insulated and have enough food stores to survive the cold months.

Scaling Up: From Hobby To Small Business

Scaling up from a hobbyist to a small business involves several key steps. Begin by developing a business plan that outlines your goals, target market, and financial projections. This plan will help guide your expansion and keep you focused on your objectives. Consider aspects like production capacity, sales channels, and marketing strategies to build your brand.

Invest in equipment and infrastructure to support your growing operation. This includes purchasing additional hives, extracting equipment, and storage facilities. Upgrading your gear will enable you to handle larger quantities of honey and manage more hives efficiently. Additionally, consider investing in a good vehicle for transporting your bees and equipment.

As you scale up, you'll need to understand legal and regulatory requirements for operating a beekeeping business. This might include obtaining licenses, meeting health and safety standards, and adhering to local zoning regulations. Building relationships with suppliers and customers will be crucial for the success of your business. Network with other beekeepers, join industry associations and explore local markets to grow your customer base.

Tips For Successful Apiary Expansion

Successful apiary expansion involves careful planning and execution. Start by expanding gradually, adding new hives in manageable increments. Rapid expansion can overwhelm you and the bees, leading to potential issues with hive management and colony health. Monitor the progress of new hives closely and make adjustments as needed to ensure they are thriving.

Invest in ongoing education and training. Beekeeping knowledge and techniques are constantly evolving, so staying informed about the latest practices, pest management strategies, and hive management techniques will help you maintain a healthy and productive apiary. Attend workshops, read industry publications, and participate in beekeeping forums to keep your skills up to date.

Implement effective pest and disease management practices. As your apiary grows, the risk of pest and disease problems can increase. Regular inspections, timely treatments, and maintaining good hive hygiene are essential to preventing outbreaks and ensuring the health of your colonies. Keeping your apiary clean and well-organized also helps in managing these risks effectively.

CHAPTER EIGHT

Pollination And Bees

The Role Of Bees In Pollination

Bees play a crucial role in the pollination process, which is essential for the reproduction of many plants. As bees collect nectar and pollen from flowers, they inadvertently transfer pollen from one flower to another. This cross-pollination is necessary for the fertilization of plants, leading to the production of fruits, seeds, and vegetables. Bees are particularly effective pollinators because they have specialized body structures, such as hairy legs and bodies, that are designed to pick up and transport pollen.

To better understand their role, imagine a bee visiting a flower. As it lands, its body comes into contact with the flower's reproductive parts, picking up pollen. When the bee moves to another flower, some of this pollen is transferred, fertilizing the flower. This process ensures that plants can produce new seeds and fruits, maintaining plant diversity and health. This symbiotic relationship between bees and flowering plants is vital for sustaining natural ecosystems and agriculture.

Beekeepers can facilitate this process by providing a diverse range of flowering plants that bloom at different times of the year. This not only supports bees in their pollination duties but also ensures that plants are adequately pollinated throughout their flowering period. Properly managing a bee colony, including ensuring they have a steady supply of nectar and pollen, helps maintain effective pollination and supports the health of the bee population.

Benefits Of Pollination For Gardens And Crops

Pollination, primarily carried out by bees, brings numerous benefits to gardens and crops. For gardeners, bees enhance the productivity of flowering plants, leading to increased yields of fruits, vegetables, and flowers. This increased productivity results in a more vibrant and bountiful garden. In commercial agriculture, bees contribute to the success of crops such as apples, almonds, cucumbers, and blueberries, significantly boosting farm profitability.

The benefits extend beyond just increased yields. Plants that are effectively pollinated tend to be healthier and more resilient to diseases and pests. This is because pollination contributes to the overall vitality of plants, enabling them to better withstand environmental stresses. In addition, the presence of bees can lead to more consistent and uniform crop production, which is essential for quality control in agricultural settings.

Gardeners can capitalize on these benefits by creating bee-friendly environments. This includes planting a variety of nectar and pollen-rich flowers, avoiding the use of harmful pesticides, and providing water sources for bees. By fostering a supportive environment for bees, gardeners can enjoy the enhanced growth and productivity of their plants, while contributing to the sustainability of local ecosystems.

Attracting Bees To Your Garden

Attracting bees to your garden is essential for optimizing pollination and supporting local bee populations. One of the simplest ways to attract bees is by planting a diverse range of flowering plants. Choose plants with varying bloom times to provide a continuous food source for bees throughout the growing season. Bees are particularly drawn to native plants, as these are well adapted to the local environment and offer the most suitable nectar and pollen.

In addition to planting flowers, consider creating bee habitats by incorporating features like bee hotels or nesting boxes. These structures provide safe places for solitary bees to lay their eggs and are especially beneficial in areas where natural nesting sites are scarce. Ensure that these habitats are placed in sunny spots and away from areas with heavy foot traffic to make them more appealing to bees.

Maintaining a chemical-free garden is also crucial in attracting bees. Pesticides and herbicides can be harmful to bees, so opt for organic gardening practices and natural pest control methods.

Providing a water source, such as a shallow birdbath with pebbles for bees to land on, can also help attract them to your garden. By creating a bee-friendly environment, you can enjoy the benefits of improved pollination and a thriving garden.

The Impact Of Beekeeping On Local Ecosystems

Beekeeping has a significant impact on local ecosystems by enhancing the health and productivity of plants and contributing to biodiversity. By maintaining healthy bee colonies, beekeepers help ensure that a wide variety of plants receive the pollination they need to thrive. This, in turn, supports the entire ecosystem, as plants are the foundation of food chains and habitats for other wildlife.

Moreover, beekeeping can help address the decline of wild bee populations. Many bee species are facing threats from habitat loss, pesticide use, and climate change. By keeping bees, beekeepers can provide important pollination services and contribute to the conservation of these vital insects. Additionally, beekeepers often engage in practices that support the broader environment, such as planting wildflowers and maintaining diverse habitats.

Beekeeping also fosters a greater awareness of environmental issues among the public. As people learn about the importance of bees and the challenges they face, they are more likely to adopt practices that support bee health and conservation. This increased awareness can lead to more widespread efforts to protect and restore habitats, reduce pesticide use, and promote sustainable agricultural practices.

Supporting Wild Bee Populations

Supporting wild bee populations is essential for maintaining biodiversity and ecosystem health. While managed honeybees play a crucial role in agriculture, wild bees also provide important pollination services and contribute to the resilience of natural ecosystems. One way to support wild bees is by creating habitats that cater to their needs. This can be done by planting native flowering plants, leaving areas of your garden undisturbed, and avoiding the use of harmful chemicals.

Providing nesting sites for wild bees is another important step. Many wild bee species are solitary and nest in ground burrows or cavities in dead wood. By leaving patches of bare soil or installing bee hotels with appropriate nesting materials, you can offer these bees a place to lay their eggs and rear their young.

Ensuring that these nesting sites are not disturbed and are located in sunny, sheltered areas will increase their attractiveness to wild bees.

Additionally, supporting local conservation efforts and organizations dedicated to protecting bee habitats can make a significant difference. Participating in community initiatives, such as planting wildflower meadows or supporting pollinator-friendly policies, helps create a more supportive environment for both wild and managed bee populations. By taking these steps, you contribute to the preservation of vital pollinator species and the overall health of the environment.

CHAPTER NINE

Advanced Beekeeping Techniques

Queen Rearing And Replacement

Introduction to Queen Rearing

Queen rearing is a crucial skill for beekeepers looking to sustain healthy colonies and manage genetic diversity. By selectively breeding queens, beekeepers can ensure strong genetics and productivity within their hives. There are several methods of queen rearing, each suited to different scales and objectives.

Methods of Queen Rearing

One common method is grafting, where larvae from a productive queen are transferred into specially prepared queen cups. These cups are placed in a queenless hive or an incubator, allowing the larvae to develop into queens under controlled conditions. Alternatively, the "walk-away split" method involves splitting a hive, leaving the original queen with one portion and allowing the other portion to raise a new queen from the existing brood.

Benefits of Queen Rearing

Queen rearing allows beekeepers to maintain strong, disease-resistant colonies. It also enables them to replace aging or underperforming queens, thereby boosting overall hive productivity and resilience against diseases and pests.

Practical Steps in Queen Rearing

1. **Preparing Queen Cups:** Select healthy larvae that are less than 24 hours old. Place them into queen cups filled with royal jelly.

2. Grafting Technique: Carefully transfer the chosen larvae into the cups using a grafting tool. Ensure minimal disturbance to the larvae during this delicate process.

3. Nurturing Queens: Place the queen cups into a queenless hive or an incubator set to the appropriate temperature and humidity. Monitor the development of the queens closely.

4. Introducing New Queens: Once the queens have matured (around 16 days), introduce them to their new hives or colonies gradually to ensure acceptance by the worker bees.

Advanced Hive Designs And Configurations

Introduction to Advanced Hive Designs

Advanced hive designs go beyond traditional Langstroth hives, offering improved management options and better protection for bees against pests and environmental stresses. These designs often focus on enhancing ventilation, insulation, and ease of inspection.

Types of Advanced Hive Designs

One innovative design is the top bar hive, which uses bars instead of frames. This promotes natural comb-building behavior among bees and allows for easier management without disturbing the entire colony. Another example is the Warre hive, designed to mimic natural bee habitats while facilitating easy vertical expansion.

Benefits of Advanced Hive Designs

Advanced hive designs provide better thermal regulation, reducing stress on bees during extreme weather. They also offer more options for natural comb management and minimize the risk of diseases associated with conventional hive designs.

Practical Implementation of Advanced Hive Designs

1. **Choosing the Right Design:** Select a design based on your climate, beekeeping goals, and management preferences. Consider factors such as insulation, ventilation, and ease of access for inspections.

2. **Constructing the Hive:** Follow detailed plans or purchase pre-built hives. Ensure materials are bee-friendly and non-toxic.

3. **Managing the Hive:** Regularly inspect and maintain the hive to prevent moisture buildup and pest infestations. Use non-invasive techniques to monitor hive health and productivity.

Organic Beekeeping Practices

Introduction to Organic Beekeeping

Organic beekeeping emphasizes natural methods and avoids synthetic chemicals to promote bee health and environmental sustainability. It involves careful hive management and a holistic approach to pest and disease control.

Principles of Organic Beekeeping

Key principles include using organic hive materials, such as untreated wood and natural waxes, and practicing integrated pest management (IPM) techniques. This includes promoting strong genetics, providing adequate nutrition, and minimizing hive disturbances.

Benefits of Organic Beekeeping

Organic practices support biodiversity and ecosystem health while producing high-quality honey and bee products. They also reduce chemical residues in honey and beeswax, ensuring a pure and natural product.

Implementing Organic Beekeeping Practices

1. **Hive Management:** Use non-toxic materials for hive construction and maintenance. Ensure hives are placed in areas free from pesticide exposure.

2. **Pest and Disease Control:** Monitor hives regularly for signs of pests and diseases. Use natural treatments such as essential oils and biological controls when necessary.

3. **Nutrition and Habitat:** Plant bee-friendly flora around the apiary to provide diverse pollen sources. Supplement with organic sugar syrup if natural forage is insufficient.

Overwintering Strategies For Bees

Introduction to Overwintering

Overwintering is crucial for bee survival during the cold months when forage is scarce. Effective strategies involve preparing hives to maintain warmth and adequate food stores throughout winter.

Methods of Overwintering

One method is insulating hives with materials like polystyrene or straw to retain heat. Another approach involves reducing hive entrances to prevent drafts while ensuring adequate ventilation to control moisture buildup.

Benefits of Effective Overwintering

Properly overwintered colonies emerge stronger in spring, ready to capitalize on early nectar flows. This reduces the need for supplemental feeding and supports colony health and productivity.

Practical Steps in Overwintering

1. **Assessing Hive Strength:** Evaluate colony size and food stores in late autumn. Ensure colonies are disease-free and have a healthy population of winter bees.

2. **Insulating Hives:** Wrap hives with insulating materials or use specially designed winter covers. Monitor hive weight and add sugar cakes or fondant if necessary.

3. **Monitoring and Maintenance:** Check hives periodically for signs of moisture or pests. Ventilate hives on mild winter days to prevent condensation buildup.

Utilizing Bee Products Beyond Honey

Introduction to Bee Products

While honey is the most well-known bee product, beekeepers can also harvest beeswax, propolis, royal jelly, and bee pollen. Each product offers unique health benefits and commercial opportunities.

Types of Bee Products

Beeswax is used in cosmetics, candles, and polishes. Propolis has antimicrobial properties and is used in natural medicines. Royal jelly and bee pollen are nutritional supplements known for their health-promoting properties.

Benefits of Diversifying Bee Product Use

Diversifying bee product use provides additional income streams for beekeepers. It also promotes sustainable beekeeping practices by maximizing the value derived from hive products.

Practical Applications of Bee Products

1. **Harvesting and Processing:** Use safe and hygienic methods to collect and process bee products. Ensure equipment is clean to maintain product purity.

2. **Marketing and Sales:** Develop niche markets for bee products based on their unique properties and benefits. Educate consumers about the health advantages of natural bee products.

3. **Regulatory Considerations:** Adhere to local regulations regarding the harvesting, processing, and sale of bee products. Ensure compliance with food safety standards where applicable.

CHAPTER TEN

Troubleshooting And Support

Common Problems And Their Solutions

Beekeeping can be a rewarding hobby, but it comes with its share of challenges. Common problems include issues with hive health, pests, and poor honey production. Identifying and addressing these problems promptly can save your hive and ensure a healthy, productive bee colony.

One frequent issue is the presence of pests such as varroa mites or wax moths. Varroa mites are tiny parasites that attach to bees and feed on their bodily fluids, weakening the colony. To manage varroa mites, use treatments such as formic acid or oxalic acid, following the manufacturer's instructions carefully. Regularly inspect your hive for signs of infestation, such as deformed or weakened bees, and ensure you are using integrated pest management techniques to keep the population under control.

Another problem is the occurrence of diseases like American Foulbrood (AFB), which affects the brood (bee larvae) and can be fatal. Symptoms include a foul odor and dark, sunken larvae. To address AFB, you need to remove and burn infected hives and frames and practice strict hygiene by disinfecting tools and equipment. Additionally, avoid using honey or bees from infected hives to prevent the spread of disease.

A common issue for beginners is poor honey production. This can be due to various factors such as insufficient foraging areas or improper hive management. Ensure your bees have access to a diverse range of flowering plants and that your hive is well-maintained.

Regularly check for adequate ventilation and correct any issues such as overcrowding or insufficient space in the hive.

When To Seek Professional Help

Despite your best efforts, you may encounter issues that require expert assistance. Knowing when to seek professional help can be crucial to the health of your hive and the success of your beekeeping endeavors.

If you encounter persistent problems that you cannot resolve with standard treatments, such as severe pest infestations or disease outbreaks, it's time to consult a professional. Beekeepers with extensive experience can offer valuable insights and solutions tailored to your specific situation. Reach out to local beekeeping associations or experienced beekeepers in your area that can provide guidance and possibly inspect your hives.

Additionally, if you are struggling with the overall management of your hives, such as complex colony behavior issues or difficulties in honey extraction, seeking professional advice can help streamline your practices. Professionals can offer hands-on training and personalized recommendations to improve your beekeeping skills and ensure your hives remain healthy.

It's also advisable to seek help if you are considering expanding your beekeeping operation. Professionals can assist with scaling up your setup, managing multiple hives, and optimizing your practices to handle a larger number of colonies efficiently.

Resources For Continued Learning

Beekeeping is a field with constant developments and evolving best practices. To stay current and improve your skills, utilize various resources dedicated to beekeeping education and support.

Books and online resources are excellent for deepening your knowledge. Look for reputable titles on beekeeping techniques, hive management, and bee health. Websites dedicated to beekeeping often offer articles, tutorials, and forums where you can ask questions and share experiences with other beekeepers.

Joining local beekeeping associations or clubs is another valuable resource. These groups often host meetings, workshops, and field days where you can learn from experienced beekeepers, exchange ideas, and gain practical knowledge. Many clubs also offer mentoring programs, pairing beginners with seasoned beekeepers for one-on-one guidance.

Online courses and webinars provide flexible learning opportunities. Many organizations offer structured programs on various aspects of beekeeping, from basic techniques to advanced hive management.

These courses can be a convenient way to enhance your skills and stay updated on new developments in the field.

Building A Beekeeping Community

Creating a network of fellow beekeepers can greatly enhance your experience and success in beekeeping. Building a supportive community helps you gain insights, share resources, and tackle challenges together.

Start by connecting with local beekeepers through associations or clubs. Attend meetings and events to meet others who share your interest in beekeeping. Participate in discussions, ask questions, and contribute your own experiences to foster a collaborative environment.

Social media platforms and online forums are also valuable for building a beekeeping community. Join groups dedicated to beekeeping where you can interact with enthusiasts from around the world. These platforms offer opportunities to learn from a diverse range of experiences and perspectives, and they often feature discussions on troubleshooting, innovations, and best practices.

Consider organizing or participating in local beekeeping events, such as hive tours or honey festivals. These gatherings provide a chance to showcase your bees, learn from others, and engage with the broader community. Building strong relationships within your beekeeping network can provide ongoing support and enhance your overall beekeeping experience.

Preparing For Long-Term Beekeeping Success

Long-term success in beekeeping requires careful planning and continuous improvement. Establishing a solid foundation and adapting to challenges will help ensure your beekeeping venture thrives for years to come.

Begin by setting clear goals for your beekeeping operation. Decide whether you want to focus on honey production, bee breeding, or pollination services. Your goals will guide your decisions on hive management, equipment, and resource allocation. Regularly review and adjust your goals as needed based on your experiences and changing circumstances.

Invest in quality equipment and maintain it properly. Regularly inspect and clean your hives, tools, and protective gear to ensure they are in good condition. Proper maintenance helps prevent problems and ensures your beekeeping setup remains effective and safe.

Stay informed about new developments in beekeeping and continually seeks opportunities for learning and improvement. Attend workshops, read industry publications, and engage with your beekeeping community to keep up with best practices and innovative techniques. Being proactive and adaptable will help you navigate challenges and achieve long-term success in your beekeeping journey.

Frequently Asked A Question And Answers

What is beekeeping?

Beekeeping, or apiculture, is the practice of maintaining bee colonies, typically in hives, to produce honey, beeswax, and other bee-related products. It also supports pollination of crops and plants.

Why is beekeeping important?

Beekeeping is crucial for pollination, which supports the growth of fruits, vegetables, and flowers. Bees are essential for the ecosystem as they help in the reproduction of many plants and crops.

What equipment do I need to start beekeeping?

Essential beekeeping equipment includes a hive (such as Langstroth, Top-Bar, or Warre), protective gear (veil, gloves, suit), a hive tool, a smoker, and a bee brush.

How do I choose the right hive for my beekeeping needs?

The choice of hive depends on your beekeeping goals, budget, and local climate. Langstroth hives are popular for their ease of management and honey production, while Top-Bar hives are often chosen for natural beekeeping practices.

How many bees are in a typical hive?

A typical beehive can contain anywhere from 10,000 to 60,000 bees, depending on the time of year and the strength of the colony.

What types of bees are in a hive?

A beehive contains three types of bees: the queen (the egg-laying female), worker bees (female bees that perform various tasks), and drones (male bees that mate with the queen).

How often should I inspect my hive?

It is recommended to inspect your hive every 7 to 14 days during the active season (spring and summer) to check for the health of the colony, the presence of diseases, and the amount of honey stored.

What is the best time to start beekeeping?

The best time to start beekeeping is in the spring or early summer, as this allows the colony to build up strength before the winter months.

How do I attract bees to my hive?

Bees are naturally attracted to new hives if they are placed in a suitable location with access to forage. You can also use attractants like lemongrass oil to draw bees to a new hive.

How do I prevent my bees from swarming?

Swarming can be minimized by ensuring the hive has enough space, regularly inspecting for signs of overcrowding, and managing hive splits or adding additional boxes as needed.

What should I feed my bees?

Bees may need supplemental feeding, especially in early spring or late fall, when natural forage is scarce.

Common feeds include sugar syrup, pollen patties, or commercially available bee food.

How do I handle a bee sting?

If stung, remove the stinger as quickly as possible, wash the area with soap and water, and apply a cold compress to reduce swelling. Over-the-counter antihistamines or pain relievers can also help manage symptoms.

What are common bee diseases and pests?

Common bee diseases include American Foulbrood, European Foulbrood, and Nosema. Pests include Varroa mites, wax moths, and small hive beetles. Regular inspections and management practices help control these issues.

How do I harvest honey from my hive?

Honey is harvested by removing the honey-filled frames from the hive, uncapping the honey cells, and extracting the honey using a honey extractor. The honey is then filtered and jarred.

How can I tell if my bees are healthy?

Healthy bees exhibit good activity around the hive, a strong brood pattern (eggs and larvae), and a sufficient amount of honey stores. Look out for signs of disease or pest infestations during hive inspections.

What should I do with my bees during winter?

In winter, ensure the hive is properly insulated and protected from harsh weather. Reduce hive entrances to prevent drafts and provide supplemental food if necessary.

Do I need a permit to keep bees?

Beekeeping regulations vary by location. Check with your local government or beekeeping association to determine if you need a permit or if there are specific rules you must follow.

Can I keep bees in an urban area?

Yes, beekeeping can be done in urban areas, but it may require adherence to local regulations and best practices to ensure the safety and comfort of your neighbors.

How do I deal with aggressive bees?

Aggressive behavior can be mitigated by ensuring proper hive management, using gentle beekeeping techniques, and maintaining strong and healthy colonies. Aggressive colonies may need to be requeened.

What are the benefits of beekeeping beyond honey production?

Beekeeping provides benefits such as enhanced pollination for gardens and crops, beeswax for various products, and a deeper connection with nature. It also offers a rewarding hobby and potential income from bee products.

CONCLUSION

Beekeeping is more than just a practice; it is a profound journey into the intricacies of nature and the harmonious relationship between humans and bees. As we conclude this exploration of beekeeping, it's clear that the benefits of keeping bees extend far beyond the mere production of honey. The practice offers a wealth of rewards, both tangible and intangible, that contribute to personal fulfillment, environmental health, and community well-being.

One of the most immediate benefits of beekeeping is the production of honey, a natural sweetener cherished for its flavor and potential health benefits. However, honey is just one aspect of what beekeeping has to offer. Beeswax, propolis, and royal jelly are valuable by-products that can be utilized in a variety of products, from cosmetics to health supplements, adding further value to the beekeeping venture.

The ecological impact of beekeeping cannot be overstated. Bees play a critical role in pollination, which is essential for the health of many ecosystems and agricultural systems. By supporting beekeeping practices, individuals contribute to the preservation of biodiversity and the sustainability of food sources. This is increasingly important in a world where pollinator populations are facing significant threats due to habitat loss, pesticide use, and climate change.

On a personal level, beekeeping offers a unique opportunity for education and connection with the natural world. The process of managing a beehive involves learning about bee behavior, hive dynamics, and the lifecycle of these remarkable insects. This knowledge not only fosters a deeper appreciation for the natural world but also encourages mindfulness and patience, as beekeepers learn to work with the rhythms and needs of their colonies.

Moreover, beekeeping can foster a sense of community. Many beekeepers find camaraderie and support within local beekeeping associations, where they can share experiences, advice, and resources. This sense of community helps build networks of support that enhance the overall experience of beekeeping.

In conclusion, beekeeping is a practice that enriches lives in multiple ways. It offers tangible rewards through honey and other bee products, contributes significantly to environmental health, and provides personal and communal benefits. As we move forward, embracing and supporting beekeeping can play a crucial role in ensuring the health of our planet and the well-being of future generations. Whether you are a seasoned beekeeper or someone contemplating taking up the practice, the journey into beekeeping promises to be both rewarding and transformative.

THE END

www.ingramcontent.com/pod-product-compliance
Lightning Source LLC
Chambersburg PA
CBHW071832210526
45479CB00001B/102